Gold Prospecting
Techniques

Rob Kanen

Kanen, Rob.

Gold prospecting : techniques.

Bibliography.

ISBN 0 9756723 5 5.

1. Prospecting - Australia. 2. Mines and mineral resources

- Australia. I. Mineral Services (Vic.). II. Title.

622.1841

Table of Contents

Figures

1.0 INTRODUCTION

Gold prospectors have won many fortunes and there are many smaller finds that have gone undocumented. Here is a general introduction with a few tips for gold prospectors and fossickers.

Gold is a native metallic element that is found in its natural form within the Earth's crust. The ancient Egyptians valued gold for its beauty and rarity, as have the many civilizations that followed. Gold is coveted for its natural beauty and resilient properties that stand the ravages of time. Throughout Human history, gold has always been amongst the most valued commodity in monetary terms. And this is why humans have travelled continents, even oceans, in search of the precious metal.

Before the 1850's, coarse gold in the form of veins, nuggets and grains, was relatively common in surface deposits. With the settling of North America, Australia and Africa by Europeans, these easily found surface deposits of gold were gradually discovered and mined until only the more subtly hidden and remote deposits are all that remain. Fortunately, new technologies and methods enable us to successfully prospect and mine these types of deposits.

2.0 EQUIPMENT

The equipment available for a prospector is varied. The equipment includes gold pans, metal detectors, dryblowers and hydraulic concentrators of various shapes and sizes. While metal detectors remain the most popular tool, knowledge of other equipment is useful, especially if the prospector or fossicker wants to expand their activities. Some of the most useful types of equipment are described here.

FIGURE 1: ACCESSORIES FOR PROSPECTING

Accessories

Accessories include the geological pick, prospecting pick, compass, times ten hand lens, safety glasses, pen knife, sample bottles and bags, hand auger (or small shovel) and gold pan. A geological pick can be used to dig holes and split rocks while metal detecting, or collect rock samples for identification or analysis. Safety glasses are used to protect the eyes when sampling or splitting rocks. A compass is necessary when prospecting away from known tracks and landmarks. A hand lens is useful for examining fine gold and minerals. The hardness of a mineral can be tested using a pen knife. A stainless steel pen knife has a hardness of 6 1/2 on Moh's hardness scale (1-10). Sample bottles and

bags are used to store fine gold and samples. They should be labeled and a list written up so locations won't be forgotten. For sampling alluvium and soils, a barrel type hand auger or small shovel is useful. A gold pan is used to separate fine gold from concentrates. A pencil magnet is used to test for magnetic minerals.

FIGURE 2: GOLD DETECTOR

Metal Detectors

V.L.F. detectors with ground balancing are the best type of detectors for prospecting. The new models of prospecting detectors have better depth and sensitivity than many of the old types. They are most useful for detecting small nuggets, which would have been missed by the old detectors. Garrett, Minelab and Whites prospecting models are popular. Garrett and Minelab detectors have had widespread success on W.A.'s goldfields.

Metal detectors will respond to any type of conductive or magnetic material. The metal detectors transmitting coil produces a primary electromagnetic field. When a conductive object encounters a primary electromagnetic field, currents flow through the surface of the object, called eddy currents, each producing its own secondary electromagnetic field. The secondary electromagnetic fields distort the primary electromagnetic field. A receiving coil within the metal detector receives the distorted primary electromagnetic field signal, ultimately producing an audio signal in response to the strengths of the secondary electromagnetic fields. Materials of greater conductivity produce larger and stronger secondary electromagnetic fields, and therefore audio signals, than smaller objects. For a given conductivity, the sizes and strengths of the secondary electromagnetic fields are controlled by the surface area facing the primary electromagnetic field rather than density or mass of the object. Larger surface areas produce larger fields and responses.

Manufacturers of metal detectors classify targets as either metal or mineral. Metal targets include all conductive, non-ferrous metals. Examples are silver, gold, copper and aluminum. Mineral targets consist of ferrous metals, magnetic minerals and conductive ground minerals. Examples of mineral targets are steel, iron, magnetite, iron oxide ground minerals and wet salt. These have lower conductivities than most metal targets. "Hot rocks", often encountered in the field, are concentrated forms of conductive iron oxide.

The audio signal produced usually varies according to the type of target. Gold tends to produce short, sharp signals while ferrous objects produce broad signals. A double blip will be produced on long thin objects, such as wire or nails. It is wise to do a bench test of different objects to determine their different responses. Some objects that can be used are a nail, silver coin and gold ring. With field experience, audio response from various targets, including "hot rocks", will become familiar so that identification will be easier.

Ground conditions affect the operation of metal detectors. Heavily mineralized and dense ground conditions cause the primary electromagnetic field to compress, resulting in loss of depth. Wet ground allows greater penetration of the primary magnetic field, providing better depth. Magnetic and conductive minerals (mostly iron oxide minerals) in ground soil produce background signals that can mask target objects. The ground cancel is used to decrease the effect of minerals in mineralized areas.

The following steps should be followed when tuning a manual ground cancelling detector; such as a Garrett or Whites prospecting metal detector: 1. Switch on and allow to stand for 10-15 minutes to stabilize batteries. 2. Check battery condition. Batteries must be in good shape for the metal detector to work properly. 3. Place tuning on automatic. 4. Place into V.L.F. mode. The V.L.F., or ground cancelling mode, should always be used for prospecting. 5. To begin with, sensitivity should be placed on half. It can be increased or decreased according to ground conditions. If it is placed on minimum, small nuggets will be missed, therefore, always place it on the maximum allowable setting. 6. Discrimination should always be at zero. 7. Adjust tuning audio so that a humming sound is barely audible. 8. Compensate for ground conditions. To compensate for ground conditions, raise and lower the search coil (from 60cm high to 15cm low). As the search coil is lowered, the audio signal will either increase or decrease in strength. If the ground cancel knob is in its midway position then it can be turned backwards or forwards to compensate for the increase or decrease in signal strength. The ground cancel knob should be moved one complete turn each time. When the audio signal remains constant as the searchcoil is raised and lowered, the ground has been compensated for. The ground cancel will have to be adjusted as ground conditions change.

Headphones should always be worn when prospecting otherwise the batteries will drain quickly. Signals are also easier to comprehend with headphones. Some prospectors have an audio boost fitted to amplify small signals, while also suppressing very loud signals. Some audio boosts also operate with a more sensitive tone. They definitely make it easier to detect small signals. Hipmounts reduce strain when detecting for long periods or using large coils.

Dryblowers

Dryblowers are mainly used to recover fine gold in areas where there are no water supplies. Nuggets are more efficiently located by using metal detectors.

A dryblower consists of a hopper/classifier overlying an inclined riffle tray fitted with an air blower. Today's dryblowers are motorized. They range in size from small, easily portable units, less than half a meter high, to large units that can process 20 tons of material per hour or more.

Dry material is fed into the vibrating hopper/classifier which removes the coarsest material. All material small enough to pass through the classifier falls into a riffle tray underneath. The heavy fraction is separated using a combination of vibration and air being blown upwards to remove the dust. Finally, heavy concentrates are removed by lifting up the riffles and sweeping the concentrates into a pan. Gold is recovered by panning the heavy concentrates.

Vibrostatic dryblowers use a combination of static electric charge, air flow and vibration to collect gold. They only differ in that the gold is precharged and later attracted to assist in retaining the gold in the riffle box. This allows damp material to be processed.

Dryshakers consist of a hopper/classifier overlying an inclined riffle tray. Dry material is placed into the hopper /classifier which remove the coarse material using vibration. All fine material passes into the underlying riffle tray. High frequency, short vibrations displace light material over the riffles and out of the tray. Gold is retained by the riffles. Air blowing is not utilized by dryshakers.

FIGURE 3: ROCKER CRADLE

Hydraulic Concentrators

When a water supply, such as a bore or stream is available, hydraulic concentrators are used to recover gold. Water can be recycled in dry areas to reduce consumption. The simplest of these is the rocker cradle. It consists of a hopper over a tray fitted with riffles, all mounted on a rocker.

FIGURE 4: GOLD PAN

Washdirt is fed into the hopper, the base of which contains small holes to prevent pebbles and boulders from passing through. The discarded coarse material should be examined for nuggets. Water is poured into the hopper and carries the fine material onto the riffle tray and over the riffles. At the same time, the cradle is rocked. Eddy currents form behind each riffle, the decrease in current velocity trapping heavy minerals. Matting covers the base of the riffle tray to help trap heavy minerals. Most gold will be trapped behind the first few riffles. The angle of decline of the riffle trays must be adjusted according to water flow and the amount and type of sediment. Too steep a decline will result in gold being washed away. A decline that is too shallow will have the riffles becoming choked in sand, preventing settling of gold. When the matting behind the riffles fills up, the riffles can be lifted and the matting removed. Finally, the heavy concentrates should be panned to remove any gold. Modern hydraulic concentrators are driven by petrol engines. Various types of hydraulic concentrators; such as, gold screws, knelson concentrators, jigs and shaking tables can be used when large amounts of washdirt are to be treated. It should be noted that all clay material must be thoroughly disaggregated before processing to prevent the formation of clay balls. Puddling and log washing machines are specially designed for this purpose.

Gold Pans and Sieves

Gold pans are made from metal or plastic. Plastic pans are easier to maintain and just as efficient as metal pans. The pan should be large (about 40cm in diameter) and contain riffles along its side to help trap gold. A pan with riffles along one half of its side is preferable to a pan with riffles along its full circumference. This allows easy collection of gold and concentrates after panning. Metal pans are often greased and should be degreased by holding over an open flame or washing in hot, soapy water.

To use a goldpan, a layer of washdirt 3/4 inch thick is placed over the base of the pan. Rest the pan in water and rake fingers back and forth to loosen and separate material. Tilt the pan and rake coarse material to the top end, letting the fines fall back. Remove this coarse material. Next, shake the pan from side to side to help the heavy minerals settle on the bottom of the pan. Repeat these two steps four or five times. Now, place the pan in water and tilt so the fine material accumulates just under the pans edge. Remove from water and tilt back, allowing a wave to form. Tip forward again, letting the wave travel forward to carry the top material out of the pan. Next, shake the pan from side to side again. Place in water and tilt, so the light material remains just under the pans edge. Remove the pan from water and tilt back, then forward, resulting in a wave carrying the top material out of the pan. The previous few steps should be repeated until only a tablespoon of fine material is left. A gentler wash action is required as the amount of wash dirt remaining decreases. Finally, swirl the remaining washdirt on the base of the pan so contents fan out and gold specks will be visible. Gold specks can be collected with a damp finger and placed in a sample bottle filled with water. A teaspoon is useful for collecting gold when large amounts are present.

Black, magnetic sands can be removed using a magnet. Place the magnet in a plastic bag so that black sands are collected on the outside of the bag. Now, the magnet can be removed and the black sands will fall away. This prevents a buildup of black sands on the magnet. A sieve is useful to initially separate coarse material from the fine fraction. Automatic gold pans, or concentrating wheels, are an alternative to manual pans for separating gold from fine alluvium and concentrates. The wheel contains riffles which pass from the pan's edge to its center. It is set at an angle so that washdirt remains at the lower edge of the wheel. Water is added to the center of the wheel by a jet spray. The circular motion and spiral action of the wheel cause gold grains to migrate towards the center of the wheel where they pass through a hole to a collecting bottle underneath. They have electric 12v motors which operate from batteries.

A new type of hydraulic concentrator, called the mini gold concentrator is replacing conventional gold pans for treating small samples of alluvium, eluvium and colluvium. The concentrator can treat twenty panloads of washdirt in the same time an expert can wash a single panload using a conventional pan. By following simple instructions a beginner can easily master the separation of gold from a shovelful of washdirt.

The unit consists of a removable dish with sieve resting on top of a lower settling pan, all clamped inside of a twenty liter bucket. To use, fill the bucket with water.

Place the entire assembly into the bucket and fasten with wingnuts. Add washdirt to the upper dish and agitate the dish in a circular motion. All large material is retained in the upper bowl by the sieve. This material is discarded by removing the upper dish and sieve. Small material passes through the sieve into the lower settling pan. Agitation washes the light material over the top of the settling pan with the help of agitation blades and a helical scraper blade. Gold and heavy minerals settle to the base of the retention bowl where they remain until panning is finished. Panning continues until the bucket fills with the discarded washdirt. Finally, the unit is removed by undoing the wingnut fasteners and the bucket is emptied then refilled to start the process over again. Any gold in the retention is removed and placed in a sample bottle. To recycle water, the entire bucket with concentrator is placed in a large container so overflowing water is collected until ready for reuse. Small in size and weighing only 3.5 kg it is easily transported.

Sieves for separating gravels and clay are made from aluminum and are available in a variety of sizes and meshes. Material is placed in the sieve and washed away by gently shaking back and forth. Determine the size of the minerals you are looking for and select a mesh to match. For instance, if searching for small stones, such as sapphires, agates or quartz, a sieve with a small -medium mesh size, such as 3-4mm may be suitable

Sample Mill

The sample mill is used to crush rock samples before testing for gold. It is powered by a petrol motor for portability. A sample is placed in the hopper which feeds the pulverizers, reducing the sample to powder. Ideally, the sample mill should be adjustable so that the desired grain size can be obtained. Most sample mills have hardened steel jaws. These can produce fine steel filings that show up in the residue when panned. When more than one sample is processed, residue from previous samples carries through. Therefore, a gold bearing sample followed by a barren sample will give positive gold results in the barren sample. When accurate results are required, the mill can be cleaned by grinding quartz between samples (sometimes, particularly with ironstone's, this is not effective).

A cheaper alternative to the sample mill is the dolly pot. A dolly pot consists of two parts: a mortar and a pestle, both of large dimensions (eg. 1 liter). It is used for crushing hand samples. Samples are broken into small pieces with a hammer, then placed in the dolly pot for crushing.

Analytical Instruments

Today, the options available to the prospector for analyzing rock and mineral samples are numerous and sophisticated. Depending on the results required, techniques such as polarized light and electron microscopy; x-ray diffraction; and chemical analysis using various spectrometric methods are available.

Polarizing and stereo microscopy are methods for identifying and examining rocks and minerals. By observing a section of a rock or mineral with a polarizing or stereo microscope the texture, structure and mineralogy of the sample can be determined. From this information, identification can be made and the origin determined. This information is of use during mining and prospecting. For routine use, lower cost alternatives are pocket microscopes and loupes.

For analyzing the composition of individual minerals emission spectroscopy (ICP) or electron microscope (microprobe) analysis is carried out. Ores containing submicroscopic gold particles within their crystal lattice are analyzed with a microprobe to determine which ores are the gold carriers and where the gold is sited.

Chemical analysis of a rock or mineral sample for gold is called assaying. For most prospectors, a low cost, moderately sensitive technique is adequate. For most gold bearing samples requiring accurate determination of the gold content, fire assaying is the most common method but not necessarily the cheapest. Modern fire assaying techniques can determine grades as low as 1g/ton and starts at prices of about $12.00 per sample. In samples containing minute trace amounts of gold, more sophisticated methods are preferred.

For the geochemical explorationist who is searching for trace amounts of gold, indicating the presence of a hidden orebody, the latest analytical techniques are almost mandatory. Atomic absorption spectrometry (AAS) , induced coupled plasma (ICP) and even mass spectrometry have detection limits in the parts per billion or less and are the preferred choice. Analytical costs are higher for these methods although bulk sampling and multi-element analysis bring the costs down.

3.0 GOLD AND ITS ORES

A mineral profitably mined for its metal content is called an ore mineral, whether it is an element, such as gold, or a compound of two or more elements, such as the sulphides and tellurides. Knowledge of the properties of gold and its ores is necessary for correct identification. This information is also necessary for selecting and controlling the mining and ore processing equipment. Visual examination of a sample is usually sufficient to reduce the number of possible identities to a few, if not a single identity. Gold is most commonly found in its elemental form, with varying amounts of silver, copper and iron as impurities but also occurs in ores; such as, the sulphides and tellurides.

Beginners sometimes experience problems when identifying gold, most commonly confusing with similar minerals; such as pyrite, chalcopyrite, pyrrhotite, pentlandite and gold colored mica. With experience, there should be no difficulty identifying gold except when it is extremely fined grained or microcrystalline. In these situations, gold cannot be easily observed and requires examination with a microscope.

The most distinctive properties of gold are its gold-yellow color, metallic luster, softness, high specific gravity and gold-yellow streak. Other minerals with a similar color and luster are often mistaken for gold. Pyrite, chalcopyrite, pyrrhotite, pentlandite and gold colored mica are the minerals usually mistaken for gold. By keeping in mind the properties of gold each of these minerals can be eliminated. Gold is the only mineral that will easily scratch, leaving a residue of gold-yellow powder. Gold is malleable while the rest are brittle, will break and flake when struck with a hammer. When fine and placed in a pan of water, gold will sink rapidly and refuse to move, the rest will sink slowly and swirl easily. Gold occurs in grains whereas mica is flaky.

Gold also occurs as microscopic and submicroscopic particles within sulphide minerals; particularly pyrite, chalcopyrite, arsenopyrite and pyrrhotite. All of these are common within veins and zones of hydrothermal alteration and replacement. They occur as macroscopic and microcrystalline grains and crystals.

Pyrite is brass-yellow in color with a metallic luster and greenish-black streak. Often, it forms perfect isometric crystals in cubic or polyhedral form.

Chalcopyrite is also brass-yellow with a metallic luster and greenish-black streak. It is easily confused with pyrite but forms tetragonal crystals instead of isometric cubes and polyhedrons. When exposed to air it often tarnishes to iridescent or deep blue. In some situations, a chemical test for copper using concentrated nitric acid may be necessary to distinguish it from pyrite.

Arsenopyrite is silver-white to steel gray with a metallic luster and grayish-black streak. When crystalline, it exhibits monoclinic crystals usually in prismatic form. When struck with a hammer arsenopyrite often gives off a garlic smell.

Pyrrhotite is brass-yellow or brownish-bronze with a metallic luster, grayish-black streak and orthorhombic crystals. Pyrrhotite is easily identified using a pencil magnet as it is distinctively magnetic.

Gold also occurs in compounds of gold and/or silver with tellurium. The tellurides, sylvanite and calaverite are mined for their gold content. They are quite rare, however, have been mined in Kalgoorlie as ores of gold.

Calaverite is brass-yellow to silver-white with a metallic luster, yellowish to greenish gray streak and monoclinic crystals that are often striated.

Sylvanite is silver-white to steel gray with a metallic luster, black streak and monoclinic crystals. The hardness of calaverite is 1 1/2 to 2 and of sylvanite 2 1/2 to 3.

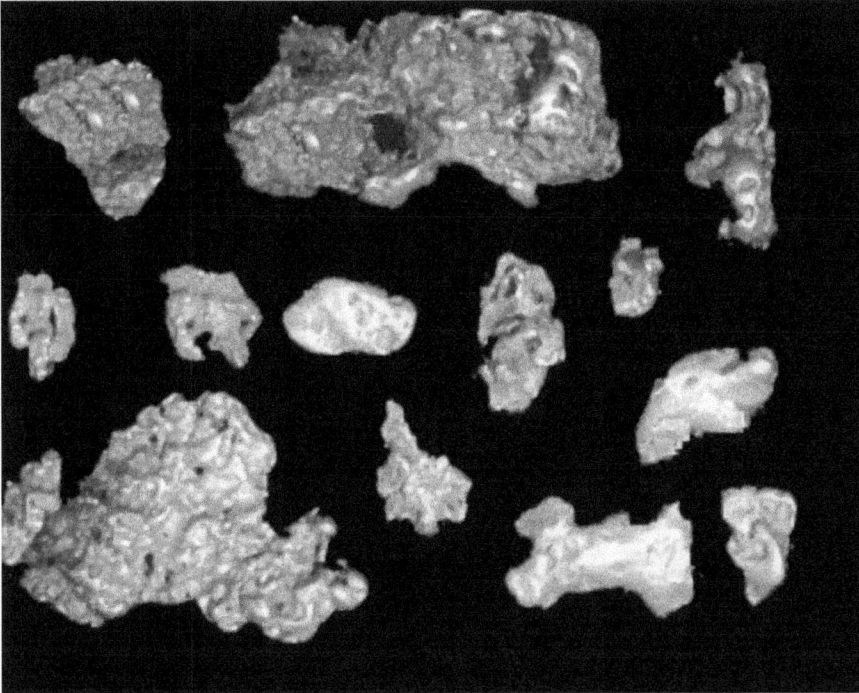

FIGURE 5: GOLD NUGGETS

GOLD Au

Color: Gold yellow to pale yellow
Luster: Metallic
Hardness: 2.5 to 3
Specific Gravity: 19.3 to 15.6
Fracture: Ductile and malleable
Streak: Gold yellow
Best Field Characteristics: Gold yellow color, high SG, gold yellow streak, softness.
Similar Minerals: Pyrite and chalcopyrite have a greenish-black streak, will sink slowly and swirl in a pan of water when fine whereas gold will sink rapidly and refuse to move. They are brittle: will break and flake when touched with a knife but won't scratch. Gold is malleable and will scratch easily. Once gold has been seen and held, future identification will be simple.

 Gold also occurs as submicroscopic particles within sulphide minerals, particularly pyrite, chalcopyrite, arsenopyrite and pyrrhotite. All of these are common within veins and zones of hydrothermal alteration and replacement. They occur as macroscopic and microcrystalline grains.

Pyrite is an iron disulphide.

PYRITE FeS2

Color: Brass yellow
Luster: Metallic
Hardness: 6 to 6.5
Specific Gravity: 4.9 to 5.2
Fracture: Uneven/brittle
Streak: Greenish-black
Crystals: Isometric, in cubes and pyritohedrons. Also occurs massive and in anhedral grains.
Best Field Characteristics: Color, streak and cubic crystal form.

Chalcopyrite is a copper iron sulphide.

CHALCOPYRITE CuFeS2

Color: Brass yellow
Luster: Metallic
Hardness: 3.5 to 4
Specific Gravity: 4.1 to 4.3
Fracture: Uneven/brittle
Streak: Greenish-black
Crystals: Tetragonal, usually massive and in anhedral grains.
Best Field Characteristics: Color and streak

Arsenopyrite is an iron arsenide sulphide.

ARSENOPYRITE FeAsS

Color: Silver white to steel gray
Luster: Metallic
Hardness: 5.5 to 6

Specific Gravity: 6 to 6.2
Fracture: Uneven/brittle
Streak: Grayish-black
Crystals: Monoclinic prismatic. Also massive and in anhedral grains.
Best Field Characteristics: Color and crystals.

Pyrrhotite is an iron sulphide with small amounts of nickel and cobalt.

PYRRHOTITE $Fe_{1-x}S$

Color: Yellowish to brownish bronze
Luster: Metallic
Hardness: 3.5 to 4
Specific Gravity: 4.6
Fracture: Uneven/brittle
Streak: Dark grayish-black
Crystals: Orthorhombic, also massive and anhedral grains.
Best Field Characteristics: Pyrrhotite is magnetic.

The tellurides are compounds of gold and/or silver with tellurium. The tellurides, sylvanite and calaverite are mined for their gold content. Calaverite is a ditelluride of gold. Sylvanite is a telluride of gold and silver. These are not common.

CALAVERITE $AuTe_2$

Color: Brass yellow to silver white
Luster: Metallic
Hardness: 2.5 to 3
Specific Gravity: 9.1 to 9.4
Fracture: Uneven/brittle
Streak: Yellowish gray
Crystals: Monoclinic prismatic with striations. Also in anhedral grains.
Best Field Characteristics: Streak and striated crystals.

SYLVANITE AuAgTe4

Color: Silver white to steel gray
Luster: Metallic
Hardness: 1.5 to 2
Specific Gravity: 8.2
Fracture: Uneven/brittle
Streak: Black
Crystals: Monoclinic prismatic. Also in anhedral grains.
Best Field Characteristics: Hardness and streak.

Ore Characteristics

Gold can be described according to its natural size and nature of occurrence. Based on these, gold occurs in six main forms:

(1) Large pieces of free gold >2mm in size that are known as nuggets.
(2) Pieces of gold and gangue (quartz, ironstone etc.) known as specimens.
(3) Coarse to fine grains of free gold 2mm to 150 microns that are visible to the naked eye.
(4) Microcrystalline gold 150 to 0.8 microns in size only visible with a microscope.
(5) Submicrocrystalline particles of gold that occur in the crystal lattice of certain sulphide ores.
(6) In compounds with tellurium.

All types show various degrees of crystallinity from rounded grains (eg. alluvial) with no crystal faces through subhedral grains with some crystal faces (hydrothermal) to crystalline grains with well developed crystal faces (hydrothermal and supergene gold). In most situations, gold is found in rounded forms, however, where open space crystallization has occurred, such as in supergene environments, crystalline gold is common.

Nuggets are well known to metal detector operators. While many nuggets are almost pure gold, impurities of iron and quartz are common. Nuggets that have been chemically deposited or altered in the weathering profile are often intergrown with ironstone.

Large grains and veinlets of gold intergrown with quartz are derived from quartz reefs and lodes and are referred to as specimens. These are also well known to metal detector operators.

Free grains of gold that are visible to the naked eye are either intergrown with gangue in primary deposits or as loose grains within secondary deposits. Machinery is required to separate gold grains from unwanted gangue. Fortunately, the high specific gravity of gold enables it to be effectively segregated and concentrated using low cost gravity methods, such as jigs, sluices, shaking tables etc.

Microcrystalline gold is common within primary deposits. Grains of gold are disseminated and intergrown within a quartz gangue or locked within sulphide minerals. Coarse grains can be liberated by crushing and grinding followed by concentration using gravity concentrators. If the ore consists of very fine grains extraction with sodium cyanide or amalgam is necessary.

Gold contained within sulphide minerals is present as small grains and particles within the crystal lattice of the mineral. Many primary deposits consist of disseminated grains of pyrite, chalcopyite, arsenopyrite and/or pyrrhotite containing significant amounts of gold and intergrown with gangue minerals. Sulphide minerals cannot be concentrated by gravity methods due to their low specific gravity. Froth flotation is common, followed by treatment with sodium cyanide to remove the gold. Such mining methods are expensive and can only be used on large deposits, however low grades can be worked.

Gold also occurs in compounds of gold and/or silver with tellurium. The tellurides, calaverite and sylvanite are mined for their gold content. They are quite rare, however, have been mined in Kalgoorlie.

4.0 GOLD ENVIRONMENTS

DEPOSIT TYPE	STYLE	DESCRIPTION	ORES
I **PRIMARY**	Quartz Reef	Large, continuous qutz reef	Free gold ∓ pyrite, arsenopyrite pyrrhotite, chalcopyrite
	Quartz Stockwork or Vein Sets	Closely spaced network of qutz veins ∓ alteration haloes around veins. At fault intersections with competent beds (e.g. meta-dolerite)	Free gold ∓ pyrite, pyrrhotite, arsenopyrite, chalcopyrite (in veins and alteration haloes)
	Shear/Fault hosted ∓ veins	Broad zones of intense deformation containing zoned disseminated alteration haloes ∓ deformed qutz veins	∓Free gold, pyrite, pyrrhotite, arsenopyrite, chalcopyrite
	Hydrothermal Replacement	Massive replacement (e.g. qutz-carbonate) in broad shear zones ∓ qutz veinlets	Arsenopyrite, pyrite, free gold ∓ pyrrhotite, chalcopyrite
II **SECONDARY**	Alluvial	Hydrodynamically accumulated grains and nuggets in seasonal creeks and gullies	Gold grains and nuggets
	Eluvial	Transported by gravity processes	Gold grains and nuggets
	Supergene	Chemically reworked, physically transported chemically precipitated gold within the weathering (laterite) profile	Gold grains and nuggets

FIGURE 6: CLASSIFICATION OF GOLD DEPOSIT STYLES

Gold occurs in alluvial, eluvial, supergene, quartz vein and stockwork, shear related and hydrothermal replacement deposits. In the general sense, alluvial refers to eluvial, colluvial, fluvial and lacustrine deposits but is restricted to the traditional meaning of stream and lake deposited gold here. Alluvial, eluvial and supergene deposits are secondary deposits formed by reworking of primary deposits. Quartz vein and stockwork, shear related and hydrothermal replacement deposits are primary deposits formed by the direct precipitation of gold from hydrothermal solutions originating in the earth's interior. Alluvial and eluvial deposits are collectively known as placer deposits. Large, continuous quartz veins are known as quartz reefs and all other large primary deposits are usually referred to as lodes. Alluvial deposits are formed by the mechanical accumulation of grains, derived from pre-existing rocks, in streams and lakes. Eluvial gold is deposited on the surface by the downward movement of material, via gravity processes, from the source which is situated above. Supergene deposits result from "in situ" weathering of mineralized bedrock which leaves behind a residue of weathered bedrock, primary and secondary ore in the weathered profile. Quartz veins are formed from hydrothermal solutions which intrude the country rock along fractures and faults. Lodes consist of a closely spaced network of quartz veins and veinlets. Shear related deposits form during shearing of the host rock along planes of stress. The associated hydrothermal solutions form gold bearing alteration haloes around the shear zones. Hydrothermal replacement deposits are formed when hot aqueous solutions react with and replace the host rock.

Alluvial Deposits

Alluvial deposits consist of hydrodynamically accumulated gold by streams and lakes. They occur on the surface, just below the surface or deeply buried. Ancient stream channels that are deeply buried are called deep leads. Gold and heavy minerals, such as magnetite, ilmenite, zircons etc. have high specific gravities; therefore, they will be transported within the base of flowing currents where they will be trapped by irregularities in the channel base or changes in current velocity. In present day channels, the heavy mineral fraction, including gold, will accumulate in pools and in cavities, fractures, depressions, behind ridges and boulders present in runs between pools. Gold will also occur in buried channel alluvium below the present river bed. Basal channel deposits will contain the most gold. These rest upon the bedrock. Other channel base deposits can occur between the surface and bedrock where they are marked by beds of coarse sediments, pebbles and conglomerates. Gold and heavy minerals will be much finer grained than the light fraction. This is due to their density and size relationships, expressed as their hydraulic ratio. Consequently, fine gold and small gold nuggets will be found with coarse sediments, pebbles and conglomerates. Another area of heavy mineral accumulation is the point bar. A point bar is formed on the inside of a bend in a meandering stream. Current flow is strongest on the outside of the bend, decreasing inwards. As a result, heavy minerals will drop out of suspension on the inside of the bend, or point bar, where current flow is least. As the stream migrates laterally, increasingly finer grained material is deposited until the channel is finally covered by fine grained alluvium. Stream channels that migrate laterally form widespread alluvial deposits that may contain gold in the abandoned channel base or point bar.

(II) ALLUVIAL

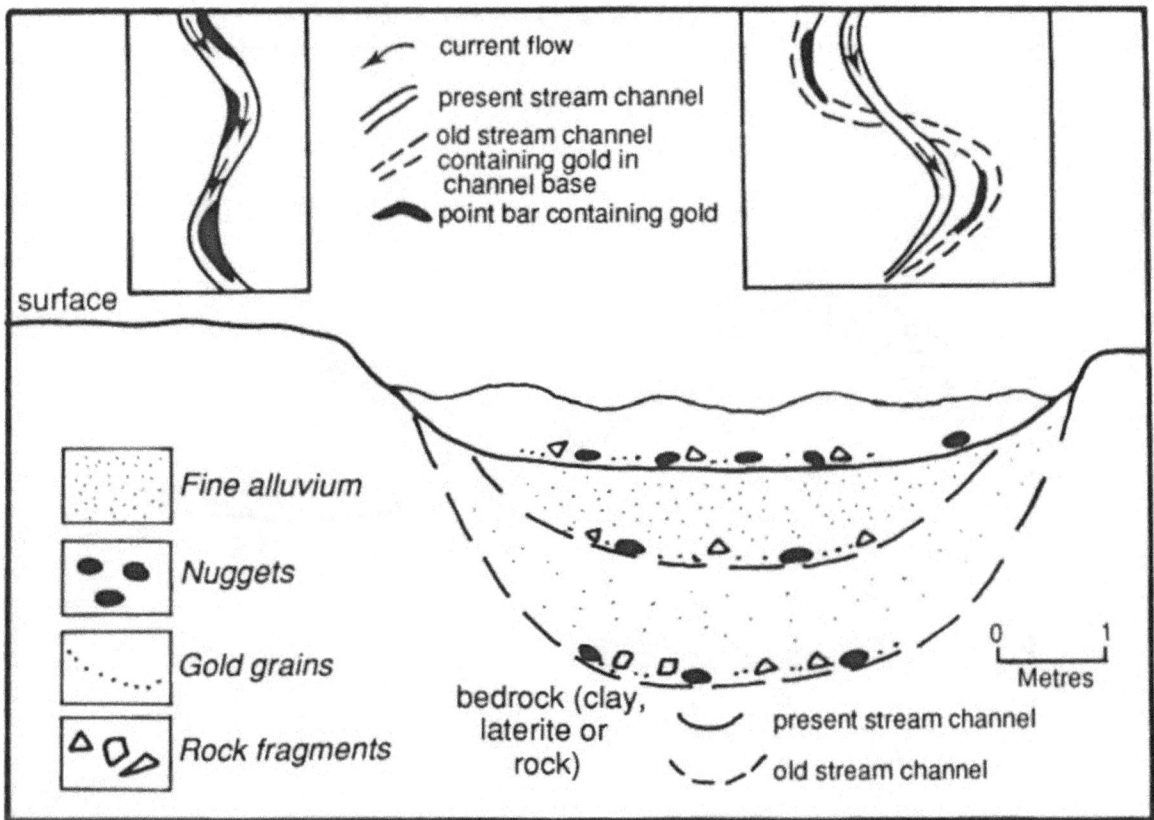

FIGURE 7: ALLUVIAL DEPOSITS

Eluvial Deposits

Eluvial gold is deposited by gravity processes on the surfaces of hills, rises and flat lying areas. Rainfall assists by carrying the surface material, or float, downslope. Eluvial deposits consist of the unconsolidated rock fragments and soil lying on the surface. It is derived from quartz reefs and other mineralized deposits (supergene, quartz reef and lode) located above. Deposits of transported material containing gold also form on the surface of hillsides where it is concentrated at changes in gradient, such as, the base of a hill. Technically, this hill wash is referred to as a colluvial deposit but is included with eluvial deposits here.

(I/II) QUARTZ REEF AND ELUVIAL DEPOSITS

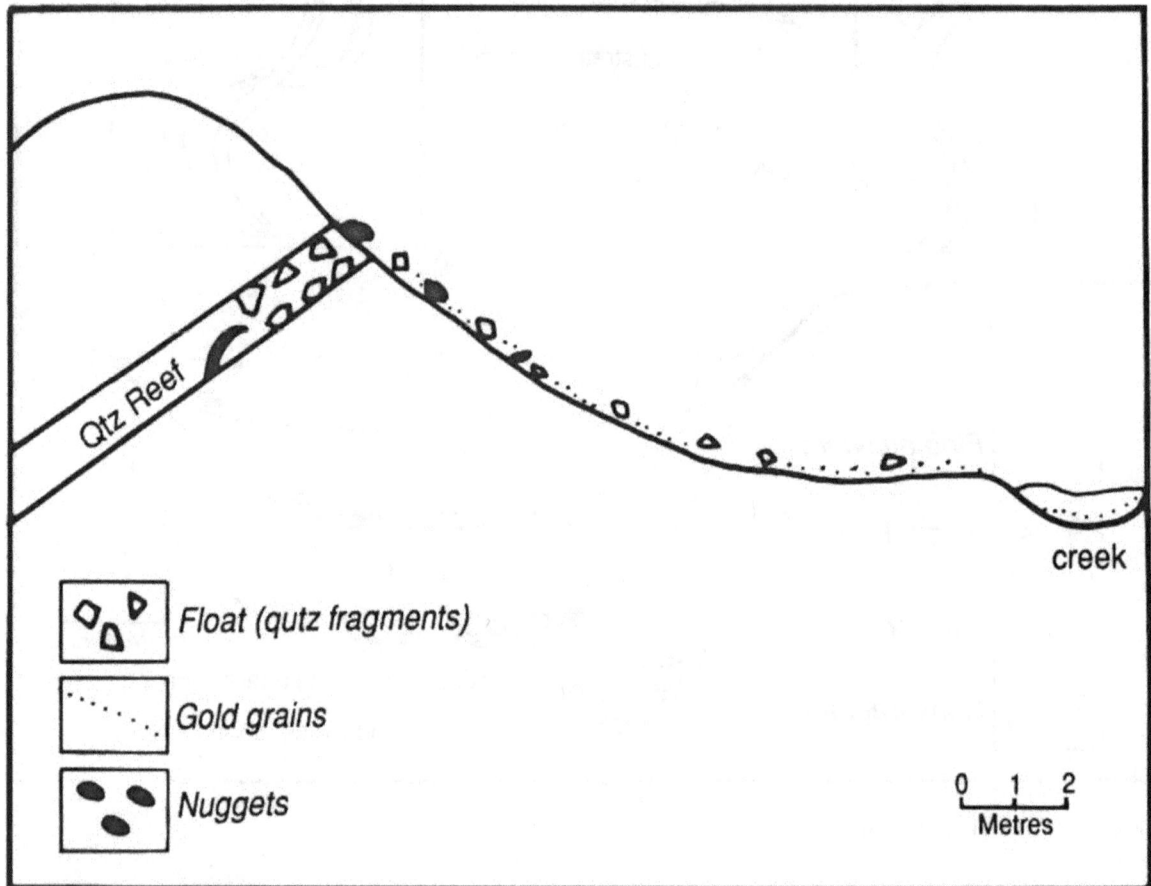

FIGURE 8: ELUVIAL DEPOSITS

(I/II) QTZ REEF AND SUPERGENE DEPOSITS

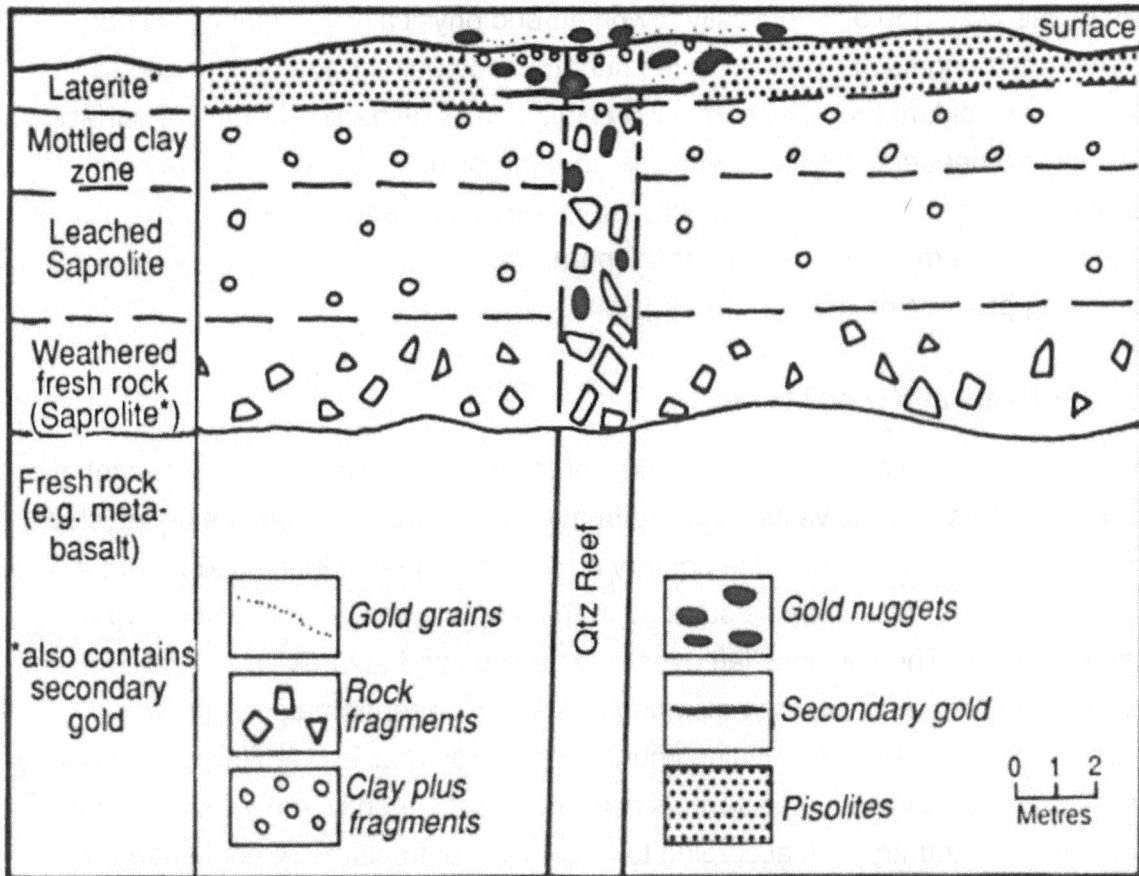

FIGURE 9: QUARTZ REEF AND SUPERGENE DEPOSITS

Supergene Deposits

Supergene deposits include both secondary and primary gold that occur in the weathering profile from "in situ" weathering of an orebody. It consists of chemically altered primary grains and nuggets, secondary grains and unaltered primary gold which may overly auriferous bedrock. Supergene gold, as it is popularly known, is the chemically precipitated gold grains and nuggets deposited within surface ironstone's, including laterite, of the weathering profile. Aqueous solutions traveling through the weathering profile transport and concentrate the gold element at or above the water table. Chemically reworked and physically transported primary grains and nuggets are present in the surface and near surface laterite and soil. Secondary gold, formed by chemical precipitation, is dispersed within the surface laterite and deeper saprolite of the weathering profile. Below the water table, unaltered primary gold, within the orebody may be present. Rich deposits, such as the "Rabbit Warren" gold find, near Leonora, have been found by the metal detecting prospector in W.A.

Quartz Reefs and Stockworks

Auriferous quartz veins and stockworks containing free gold are keenly sought after by prospectors. Quartz veins originate from hydrothermal solutions being injected along fractures and faults in the country rock. The source of these hydrothermal solutions varies. They may be sourced from rising magmas that crystallize to form igneous rocks. The solutions left over are injected into fractures and faults overlying the igneous bodies. They may also originate from a deeper magma source or metamorphism of the surrounding country rock. Fractures and faults cut the country rock at various angles and in various patterns. Consequently, the infilling quartz veins cut the country rock according to the pattern of fractures. A concentrated network of gold bearing quartz veins forms quartz stockwork deposits. Widely spaced networks of quartz veins are known as vein sets. Saddle reefs form when quartz veins are concentrated in the apex of an anticline.

Quartz veins are classified as hypothermal (high temperature), mesothermal (medium temperature) or epithermal (low temperature) veins. Hypothermal veins are deposited at great depths (>3600m). Epithermal veins are deposited near the surface (Gold is not only present within the quartz vein itself but also in the altered zone of wall rock associated with quartz veins. Gold occurs as free grains in quartz veins and submicroscopic particles within sulphide minerals. The auriferous sulphide minerals are concentrated in the altered zone of wall rock adjacent to quartz veins and within the quartz veins themselves.

(I) QTZ STOCKWORK AND VEIN SETS

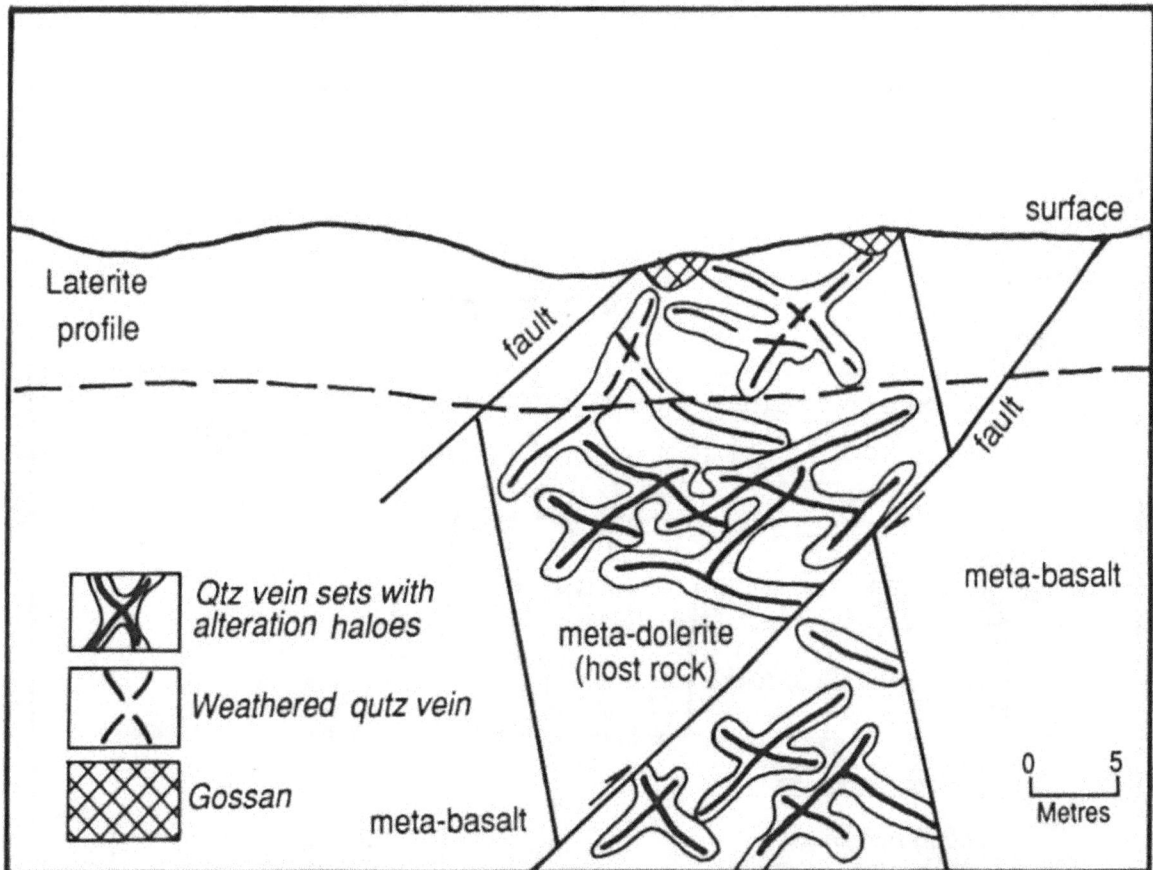

FIGURE 10: STOCKWORK DEPOSITS

In the Yilgarn Block, most auriferous quartz veins are contained within mafic rock types (particularly meta-basalts, meta-dolerites, amphibolites) within volcanic dominated greenstone belts. Ultramafics and felsic volcanics also contain gold deposits (in fact, all rock types are represented). Auriferous quartz veins are mainly controlled by shear zones and faults, particularly where faults cut competent (brittle) beds, such as dolerite, contained within less competent country rock. Vein type mineralization occurs at Kalgoorlie, Leonora, Wiluna, Cue, Mt. Magnet, Sandstone, Marble Bar etc.

(I) MASSIVE REPLACEMENT

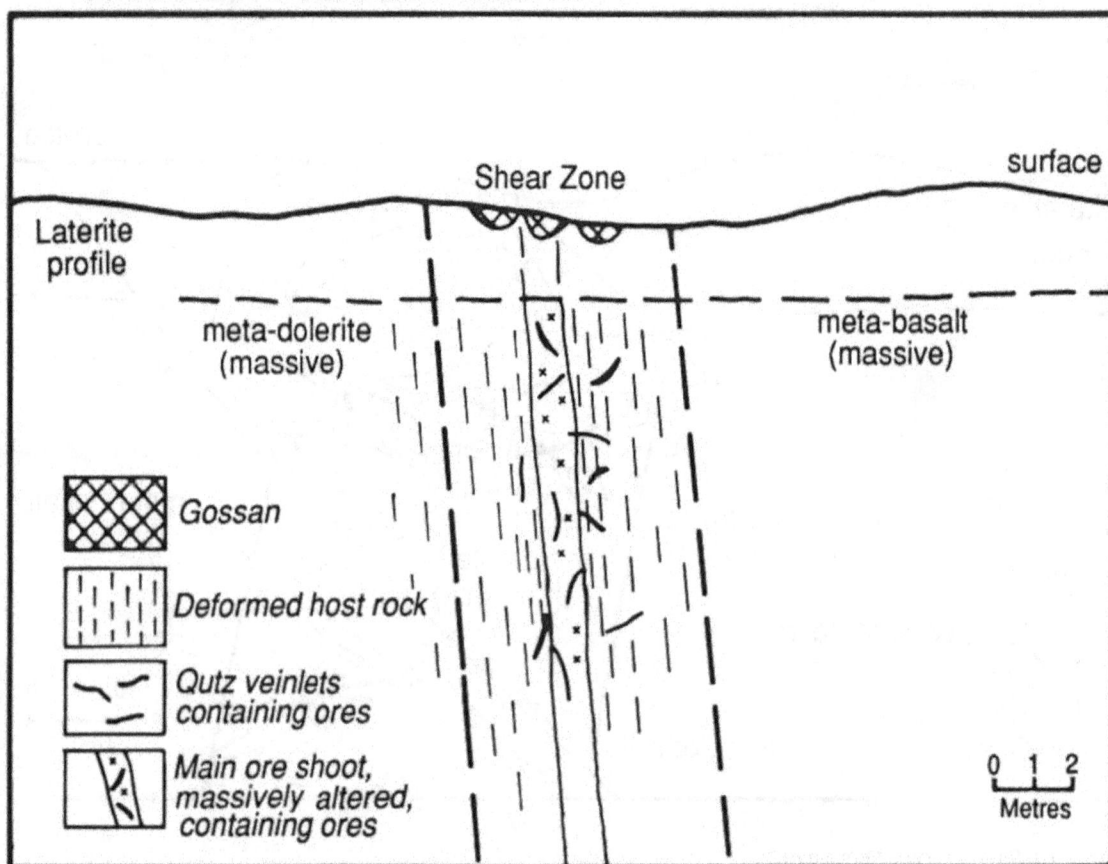

FIGURE 11: SHEAR HOSTED DEPOSITS

Other

Shear related, Banded Iron Formation hosted and hydrothermal replacement deposits also occur (listed in decreasing abundance). Shear related gold mineralization consists of alteration haloes (a form of replacement) around zones of intense deformation (shear zones), formed from the reaction of hydrothermal solutions with the wall rock. Gold is present as submicroscopic particles within sulphide minerals that occupy the alteration haloes. Quartz veining can also be present.

B.I.F. (Banded Iron Formation) hosted deposits are an example of host rock control, being restricted to a B.I.F. unit. They contain either replacement style or auriferous quartz vein mineralization. In replacement style B.I.F. deposits, hydrothermal solutions transport the gold element along faults, forming auriferous deposits by replacing magnetite and carbonates within B.I.F. At Hill 50, near Mt. Magnet, gold is concentrated along northeasterly trending faults cutting the Banded Iron Formation. Gold is present as submicroscopic particles within sulphide minerals plus/minus free grains. The sulphide minerals replace carbonates and magnetite within B.I.F. Auriferous quartz veins, within B.I.F., occur in the same fashion as those described under Quartz Reefs and Stockworks. These deposits are entirely restricted to a host B.I.F. unit.

With hydrothermal replacement deposits, hydrothermal solutions react with and replace the host rock, forming massive or disseminated gold deposits. In the massive style these typically preferentially replace a specific bed. This style is called stratabound as it is restricted to a single bed, or stratum. These can occur in combination with the deposit styles described above.

5.0 PROSPECTING METHODS

In the early days, prospectors adapted their equipment to environmental conditions so that dryblowers were used in dry areas and hydraulic concentrators in wet areas. Today, metal detectors have superseded the dryblower as the major prospecting tool. The gold pan and sample mill also have their uses.

Metal Detecting

The abundance of iron oxides on the surface of goldfields caused many problems for the first metal detectors. This led to the introduction of ground cancelling machines in 1975. They proved effective and became popular, although there are still areas where ground cancelling machines cannot operate.

The metal detecting prospector is concerned with alluvial, eluvial, and supergene gold. Areas that have been surface worked by the early prospectors mark surface gold producing districts. Many nuggets have been found on and adjacent to these worked patches. Together with the geology, they should be regarded as initial guides to metal detecting areas.

Alluvial gold can be found in the small seasonal streams that cut these areas. Basal channel deposits concentrate heavy minerals and are the most prospective deposits. Laterally migrating streams that change course regularly will contain gold in the abandoned channel base and point bar. These deposits will occur in the present day stream channel and immediately adjacent ground.

Eluvial gold can be found on low hills, rises and flat lying areas adjacent to the above locations. These are often covered with quartz and ironstone rubble. Eluvial deposits are concentrated at a change in gradient, such as the base of a hill.

Supergene deposits are found on low hills or flat lying areas that have developed laterite profiles over bedrock. The occurrence of supergene gold is difficult to predict since it is controlled by a complex combination of processes. It is generally present above weathered orebodies where it is concentrated and deposited by certain solutions traveling through the weathered zone. Secondary gold occurs in the surface laterite and deeper saprolite of the weathered zone (laterite profile) and consists of dispersed crystalline grains. Chemically altered and physically transported primary grains and nuggets, derived from the original orebody, occur in the surface and near surface with the secondary deposits. These are the main targets for metal detector operators. Weathered bedrock is also often covered by thick sequences of transported overburden (sand sheets, alluvium and colluvium). This material should be avoided as it has been diluted and mixed. The prospector should also beware of laterite profiles developed over alluvium and colluvium instead of bedrock.

In most situations, alluvial, eluvial and supergene deposits will only form over bedrock or residual laterite profiles. Exceptions to this occur when alluvial and eluvial systems are fed from these areas or where deeply buried ancient river channels exist.

The beginner should locate ground that is not heavily contaminated by iron oxide or ironstone nodules that play havoc with the detectors audio. Even so, the ground cancel will have to be adjusted as the prospector moves over new ground. Audio drift or badly erratic audio signifies that the ground cancel needs adjusting. If the ground cannot be compensated for the prospector should move to a new area.

"Hot rocks" are always encountered by the prospector. These are concentrated forms of magnetic or conductive iron oxide that behave in a similar fashion to gold. Mostly, they will give broad signals. To test whether a "hot rock" contains appreciable amounts of gold, switch to the ferrous target identification mode of your detector. With other detector types that do not have a ferrous target identification mode, the hot rock can be cracked open and both halves tested. If both halves give the same response, it can be discarded. Of course, it may not contain any gold; it may just be a lump of iron oxide.

Gridding is employed to comprehensively cover a section of ground. After a nugget has been found, the area should be gridded and explored thoroughly. This is done by marking a rectangular grid with a pick or trailing a chain. A grid is formed by marking the corners of a 10m by 5m rectangle. Next, the ends of the rectangle are marked off in one step (1m) intervals. Detecting is started at one corner and continues along the length of the rectangle. When this is completed, the operator moves to the next grid mark and follows this lengthwise so that he eventually moves across the whole of the rectangle in 1m intervals. Even when an area is gridded it is possible to miss gold. The best solution is to slow down and detect carefully.

Dryblowing and Hydraulic Concentrating

Dryblowers and hydraulic concentrators are used to recover fine gold and nuggets. Consequently, alluvial, eluvial and supergene deposits, which are most likely to concentrate fine gold, are the main targets.

Alluvial deposits are restricted to present day stream channels and immediately adjacent ground. The latter is deposited by migrating stream channels that change course regularly (being deposited in the abandoned channels). Basal channel deposits usually contain most of the gold. These are marked by conglomeratic or coarse grained beds in the subsurface or along deeply cut banks. Places to look for alluvial gold include creeks and gullies along hill sides and in depressions between hills. Eluvial deposits occur on hillsides and in depressions between hills.

Loaming

Loaming is the technique of systematically sampling and testing soil for particles of gold. Loaming is carried out to locate and test gold deposits and trace shows back to their source. Loaming using gold pans was widely employed by the early prospectors. Today, sampling machines can be used instead of gold pans to test soil samples for gold. Automatic gold pans (concentrating wheels) and small, portable dryblowers are two examples.

Prospecting for Quartz Reefs and Other Deposits

Reef prospecting involves locating gold bearing quartz veins. Most of the accessible reefs have probably been found by early prospectors and explorationists; consequently, remote and poorly outcropping reefs are more likely to be found. Today, in the short term, this form of prospecting is not as rewarding as metal detecting.

Surface weathering of outcropping quartz reefs distributes gold away and downslope from the reef, resulting in the formation of alluvial and eluvial deposits. Consequently, it is possible to trace the alluvial or eluvial deposit upstream and upslope until the source reef is located. Often, the reef has been completely weathered away, leaving only alluvial and eluvial deposits.

Once a quartz reef is located, it may be rewarding to follow the reef along its length searching for auriferous locations. Gold concentrations can increase and decrease along the length of a quartz reef.

In areas that are poorly exposed, reef prospecting is mainly restricted to the low hills and rises, where outcrop is best. In deeply weathered areas, the surface expression of quartz reefs will be in the form of supergene deposits (described previously). The presence of gossan is an indicator to an underlying orebody. Gossan is the weathered product of an orebody and is stained various colors from the oxidation of ore minerals. It generally consists of iron oxide minerals with a relict box work texture left behind after the removal of cubic pyrite. Since pyrite is often associated with gold deposits, gossan may indicate the presence of an orebody.

Within greenstone belts, mafic rock types should be targeted as the most likely host rocks. Meta-basalts and meta-dolerites are common host rocks; however, virtually all rock types are represented. Auriferous quartz veins are mainly controlled by faults and shear zones. The major regional faults and shears are barren of gold mineralization. Secondary (and later) faults and shears, leading off the regional structures, contain major quartz reef and lode deposits. Alteration haloes around quartz veins and structures (faults, shears and fractures) are indicators to gold mineralization (particularly the presence of iron sulphide minerals). Gold is present as submicroscopic particles in sulphide minerals (pyrite, pyrrhotite, chalcopyrite, arsenopyrite) plus/minus free grains in veins. The best method for correctly identifying sulphide minerals, particularly microcrystalline grains, is polarized light microscopy (petrography).

Whenever quartz veins or zones of alteration are encountered in the appropriate geological environment they should be sampled. In some cases, fresh bedrock will not be preserved in outcrop. Laterites, the weathered product of fresh rock, are most common. In some situations, it is sufficient to sample laterite, provided the laterite profile is residual (overlying bedrock) and unmodified, since gold is fairly chemically immobile and resistant to chemical weathering, some residual gold will usually be preserved. This will vary from area to area according to the degree and type of weathering. One disadvantage is that the original rock texture is obscured by weathering; therefore the prospector cannot be certain of the rock type being sampled. Once the sample is obtained, a sample mill or dolly pot is required to crush the sample. The sample can then be panned to determine whether any free gold is present; or preferably, samples can be assayed by a lab (this would not be of interest to the small scale prospector). If the sample gives a significant result, it can then be examined microscopically to determine the nature of the ore (whether as free gold grains or in specific sulphide minerals).

Geochemical Prospecting

Prospectors with some vision and adequate resources prefer geochemical testing of soils and rocks to the loaming technique. Geochemical sampling can identify and locate deposits with poor surface signatures, such as, when gold particles are present in insufficient quantities or coarseness to show up in a gold pan or concentrator. Geochemical prospecting is carried out to locate hidden orebodies that are without visible surface indications or to define the location, distribution and size of a known deposit. With this type of prospecting, samples are collected and sent to a lab where they are analyzed for gold and elements associated with gold (pathfinders, particularly As). This type of prospecting can identify a variety of deposits- quartz reef and lode, supergene, hydrothermal replacement etc. Soil sampling is done to locate and analyze the distribution of alluvial and eluvial deposits or locate anomalies that overlie hidden orebodies. Geochemical sampling of outcrops can be done to determine their gold content. Soil sampling or stream sediment sampling can be carried out to analyze gold or pathfinder elements. For detailed evaluation of prospects, contour maps can be drawn to show the distribution of elements. These may show the distribution of alluvial and eluvial deposits or the location of anomalies, indicating the presence of an orebody (where elements are most concentrated): for example, a reef.

Geochemical sampling permits accurate estimation of the grades and reserves of a gold deposit.

For some deposits containing microscopic gold (some shear related, hydrothermal replacement, quartz reef and stockwork deposits) geochemical analysis is the only method able to identify them.

6.0 SOME GOLD PROSPECTING LOCATIONS

Western Australia

The following locations in the Murchison and Eastern Goldfields are all proven gold producers. Some locations are renowned as secondary (nugget) gold producing areas while others are mainly known as primary gold producing areas. Nuggets have been found with metal detectors in all the major Western Australian gold fields. In this list, each area is defined according to the geological unit it resides within and the types of gold deposits are described.

Southern Cross Area

Gold mineralization is restricted to the Southern Cross Greenstone Belt and smaller greenstone belts adjacent to this. The Southern Cross area is dominated by primary gold production. B.I.F. hosted and shear hosted deposits are the main primary types, however, quartz reef deposits also occur. The Southern Cross area has poor prospectivity for secondary gold and few nuggets have been found in this region.

Coolgardie Area

Part of the Norseman-Wiluna Greenstone Belt, Coolgardie was the site of the first major gold find within W.A. in 1892. This was at Bayley's Reward Lease where over 5000oz of gold was initially found over a quartz reef. The Coolgardie area has produced many nuggets and also contains primary gold deposits at Gibraltar, Bullabulling, Bonnie Vale and Bayley's.

Kalgoorlie Area

Part of the Norseman-Wiluna Greenstone Belt. Kalgoorlie was the site of rich alluvial gold finds at Mt. Charlotte in 1893. Major primary deposits were found nearby. The

Kalgoorlie-Boulder Golden Mile was the richest mile in the world during its heyday. Gold mining is still carried out here today. The Golden Mile deposits are primary deposits that are composed of quartz reefs and stockworks within the Golden Mile Dolerite. The associated alluvial deposits have been worked out. Many primary and secondary deposits occur in the surrounding region. Nuggets have been recovered at Hogans, Kanowna, Black Hills and Bulong among other places.

Kambalda Area

Part of the Norseman-Wiluna Greenstone Belt. The area around Kambalda contains many primary and secondary deposits. Larkinville is famous as the site of W.A.'s largest gold nugget, called the Golden Eagle. The Golden Eagle weighed 1175 oz and was found in 1932. Numerous other large nuggets have been found here.

Norseman

At the southern tip of the Norseman-Wiluna Greenstone Belt. Several large primary deposits occur at Norseman and nearby. These are all shear hosted or quartz reef deposits. Some secondary gold deposits occur at Norseman and Dundas, however, secondary gold and nuggets have not gained prominence in this area.

Broad Arrow Area

Part of the Norseman-Wiluna Greenstone Belt. Numerous large primary and secondary gold deposits occur in the district. Black Flag, Broad Arrow, Lady Bountiful and Ora Banda are major primary deposits. Secondary gold, including nuggets have been found at Bardoc, Broad Arrow and Paddington.

Leonora Area

In the middle section of the Norseman-Wiluna Greenstone Belt. Large primary gold deposits occur at Sons of Gwalia, Harbour Lights, Malcom and other places. These are shear hosted and quartz reef deposits. Major secondary deposits have been worked at the Specking Patch, Lake Darlot and the Goana Patch. Nugget finds have been made at all the secondary localities.

Laverton Area

Part of the Norseman-Wiluna Greenstone Belt. Many primary and secondary gold deposits occur in the district. Quartz reef, shear hosted and hydrothermal replacement deposits are at Murrin Murrin, Ida H, Euro, Westralia and Lancefield. Large secondary gold finds have been made at Red Flag, Murrin Murrin, The Patch, Laverton and Erlistoun.

Agnew Area

Within the Norseman-Wiluna Greenstone Belt. Several large primary and secondary gold deposits occur in the region. Quartz reef and shear hosted deposits are present at Great Eastern and McCaffery's. Large secondary deposits have been worked at Lawlers, Sir Samual and Kathleen Valley.

Wiluna Area

Part of the Norseman-Wiluna Greenstone Belt. Large quartz stockwork and reef deposits are present at Wiluna. Secondary gold deposits also occur at Wiluna.

Mt. Margaret Area

At the Southern tip of the Mt. Magnet-Meekatharra Greenstone Belt. Numerous primary and secondary deposits have been worked in the region. B.I.F. hosted and quartz reef deposits are present at Hill 50 and Watertank Hill. Secondary deposits have been worked in and around Mt. Magnet.

Sandstone Area

Primary and secondary deposits occur at Sandstone and within a radius of 50km in the Sandstone Greenstone Belt. Large nugget finds have been made here.

Cue Area

Part of the Mt. Magnet-Meekatharra Greenstone Belt. Shear hosted, quartz reef and stockwork deposits have been worked at Big Bell, Golden Crown, Great Fingall and Cue. Numerous secondary deposits have been worked at Cue and in the surrounding region, particularly at Day Dawn, The Island and Tuckabianna.

Meekatharra

At the Northern tip of the Mt. Magnet-Meekatharra Greenstone Belt. Many primary and secondary gold deposits occur in the district. Shear hosted, quartz reef and stockworks form major deposits at Reedy's, Nannine, Kohinoor, Kiorara, New Alliance etc. Secondary gold, including nuggets, was recovered at Meekatharra, Nannine, Yaloginda, Ruby Well, Hohens Find etc.

Victoria

Central Victorian Goldfields

Dunolly/Rheola/Kingower

The Dunolly area is old reliable as far as gold fossicking is concerned. Thousands of fossickers have ambled through the state forests around Dunolly and much gold has been found here.

Daylesford/Blackwood

You may find color in the creeks around Daylesford and Blackwood.

Bendigo

Probably not much left here. Try the old alluvial diggings in the state forest.

Wedderburn

The local prospecting shop may be able to help you here. Try the old diggings.

Avoca

Try the old diggings in the state forest.

Northeast Victoria

Woods Point/Walhalla

Likely to be some alluvial gold in the creeks and rivers. Have a go with a pan or sluice.

Tallangalook/Mansfield

Try the old workings in the creek at Tallangalook.

Beechworth/Eldorado

May be some alluvial gold in the creeks and rivers.

Omeo

There are old deep lead workings in the cliffs and hills around Omeo.

7.0 GLOSSARY

Adamellite: A variety of granite containing >2/3 potassium feldspar compared to plagioclase (in addition to other granite minerals).

Alluvial: In the traditional sense, formed by the hydraulic accumulation of grains from pre-existing rock in rivers, streams and lakes. In the modern sense, it includes eluvial and colluvial deposits in addition to hydrodynamically accumulated gold.

Amphibolite: A metamorphic rock consisting of amphiboles and plagioclase.

Anticline: An inverted "U" shaped fold.

Aphanitic: Grains not visible with the naked eye.

Archean: Greater than 2500 m.y. old. Rocks of this age contain important non-precious and precious metal deposits.

Auriferous: Containing gold.

Banded Iron Formation: A banded rock consisting of alternating bands of iron oxide (e.g. magnetite or hematite) with white chert.

Basalt: A mafic, or dark colored, aphanitic and porphyritic igneous rock consisting of plagioclase feldspar, pyroxenes, amphiboles and olivine. It is extrusive, flowing out onto the earth's surface.

Basin: A large depression on the earth's surface which sediments accumulate in.

Chert: A sedimentary rock consisting of microcrystalline quartz, formed by precipitation in sea water or accumulation of organic matter.

Claystone: A very fine grained (grains <.003mm in diameter) sedimentary rock.

Colluvium: Surface material transported by gravity.

Conglomerate: A sedimentary rock containing rounded pebbles and cobbles.

Country Rock: The rock surrounding and containing intrusive bodies, such as veins and dykes.

Cryptocrystalline: Submicrocrystalline. Crystals only identifiable with x-rays.

Deep Lead: A deeply buried ancient river channel.

Dip: The angle of decline of a bedding plane which is at right angles to the strike.

Disseminated: Sparsely distributed minerals through the host rock.

Dolerite: A fine grained, even granular, mafic, intrusive igneous rock. Contains plagioclase, pyroxenes, quartz, hornblende and olivine.

Dyke: A vertical sheet like intrusive body.

Eluvial: The unconsolidated rock fragments and soil deposited by gravity processes on the surface.

Fault: A plane of weakness or movement in the earth's crust.

Felsic: A general term to describe an igneous rock containing dominantly light colored minerals, such as, quartz and feldspar.

Float: Rocks and minerals on the surface of hills, carried downslope by gravity processes.

Fracturing: The breaking up of rock, forming many breaks.

Gabbro: A coarse grained, even granular, mafic igneous rock containing plagioclase, hornblende, biotite, pyroxenes, and quartz.

Gibber Plain: A level plain covered with pebbles.

Gneiss: A coarse grained, foliated rock consisting predominantly of quartz, feldspar, biotite and hornblende. Platy biotites form the foliation. Formed by high grade regional metamorphism.

Gossan: The rock, rich in iron oxide minerals (with or without ore minerals) left from weathering of an orebody. Often forms a boxwork texture from weathering of cubic pyrite and galena. Also stained various colors from oxidation of ore minerals.

Granite: A coarse grained, even granular, felsic igneous rock containing mainly quartz, feldspars, mica and hornblende.

Granodiorite: A variety of granite containing >2/3 plagioclase compared to potassium feldspar (in addition to other granite minerals).

Granulite: A metamorphic rock containing feldspars, pyroxenes and garnet.

Greenstone: A term applied to slightly metamorphosed basalts. Used generally to describe metamorphosed rocks in linear belts of Archean age.

Hardcap: The resistant cap to hills usually in the form of gold bearing cemented conglomerates.

Host Rock: The parent rock which is replaced by secondary mineral solutions.

Hydrothermal Replacement: The replacement of pre-existing rock by hot, aqueous (hydrothermal) solutions.

Igneous: A rock formed by the cooling of magma originating from the earths interior.

Ironstone: In one instance, the iron oxide rich rock left from weathering of the parent rock.

Laterite: The surface rock, rich in iron oxide minerals, left from weathering of the parent rock.

Laterite Profile: A profile of weathered rock formed by surface weathering. The ideal laterite profile consists of several horizons, beginning with saprolite at the base (overlying fresh rock) and finishing with laterite on the surface.

Leader: A medium size quartz vein containing gold.

Lithology: Rock type.

Lode: A large primary gold deposit.

Mafic: An igneous rock that contains predominantly dark colored, ferro-magnesium minerals, such as hornblende, pyroxenes, olivine and biotite.

Meandering: A high sinuosity, or curvy, river channel.

Meta-: A prefix used to indicate a slightly metamorphosed rock retaining its original texture.

Metamorphism: Alteration of a rock due to an increase in temperature and/or pressure.

Mineral: A homogenous solid of a given composition that is formed by natural inorganic processes. Components of a rock.

Nugget: A free piece of gold, large in size.

Ore Mineral: A mineral mined for its metal content.

Point Bar: The bar formed on the interior of bends in meandering rivers or streams.

Porphyritic: Containing large crystals (phenocrysts) in an aphanitic, or glassy, ground mass. A term applied to volcanic rocks.

Primary Gold Deposit: A deposit formed by the direct precipitation of gold from hydrothermal solutions originating in the earth's interior.

Quartz: A crystalline tectosilicate consisting of SiO_2.

Quartzite: A rock consisting entirely of quartz grains.

Quartz Reef: A large, continuous auriferous quartz vein.

Rhyolite: An aphanitic, porphyritic, volcanic extrusive rock containing quartz, biotite, feldspar and hornblende.

Sandstone: A fine to coarse grained sedimentary rock formed by the accumulation of grains from pre-existing rock. Grain size >.062mm to <2mm.

Schist: A foliated metamorphic rock. Foliations comprise more than 50% of the rock. Usually consists of micas, actinolite, tremolite, talc, chlorite or serpentine.

Secondary Gold Deposit: A deposit formed by the reworking of a primary gold deposit including chemical and physical processes, such as, very low temperature hydrothermal solutions containing acids; rain, temperature and wind. Alluvial, eluvial and supergene gold deposits are secondary deposits.

Sedimentary: A rock formed by the accumulation of grains from pre-existing rock, organic processes or precipitation.

Shear Zone: A planar zone of ductile or brittle/ductile deformation.

Siltstone: A very fine grained sedimentary rock. Grain size >.0039mm and <.062mm.

Stratigraphy: The study of stratified rocks, predominantly the character, age and correlation of strata.

Strike: The direction, or trend, of a bedding plane.

Stringer: A small quartz vein containing gold.

Supergene Deposit: The primary and secondary gold left in the oxidized weathering profile from weathering of an orebody.

Supergene Enrichment: Secondary enrichment of minerals at or above the water table due to low temperature, aqueous solutions traveling through the weathering profile.

Syncline: A "U" shaped fold.

Tectonic Unit: A structural unit of the crust.

Telluride: The compound of one or more elements, such as gold or silver, with tellurium.

Ultramafic: An igneous rock consisting almost entirely of mafic, or dark colored, minerals. The common ultramafic minerals are olivine, pyroxenes, and amphiboles.

Volcanic: An aphanitic igneous rock intruded near to, or extruded onto, the earth's surface.

Weathering: The removal and alteration of minerals due to the surface processes of rain, temperature and wind.

8.0 References

Geochemical Prospecting:

Smith, R.E. (ed.), 1982. "Geochemical Exploration in Deeply Weathered Terrain". CSIRO Institute of Energy and Earth Resources, Div. of Mineralogy, Floreat Park, W.A.

Geology of W.A. Gold Deposits:

Ho, S.E. and Groves, D.I. (eds.), 1987. "Advances in Understanding Precambrian Gold Deposits". Geol Dept. and Univ. Ext., Univ. of West Aust., Publ. 11.

"   "   1988. "Recent Advances in Understanding Precambrian Gold Deposits Vol. 2". Geol Dept. and Univ. Ext., Univ. of West Aust., Publ. 12, 360p.

"   "   and Bennet, J.M. (eds), 1990. "Gold Deposits of the Archean Yilgarn Block: Nature, Genesis and Exploration Guides". Geol. Dept. and Univ. Ext., Univ. of West Aust., Publ. 20, 407p.

Metal Detecting:

Stone, D, Sargent, R. and Stone, S: Metal Detecting for Gold and Relics in Australia. Outdoor Press, Burwood, Vic.

Lagal, R. and Garrett, C., 1980. "The Complete V.L.F./T.R. Metal Detector Handbook". Ram Publ. Co., Dallas, Texas.

Prospecting Locations:

Johnnston, T.E., 1980. "A Guide to Alluvial and Dryblowing Patches in the Eastern Goldfields". T.E. Johnston and Associates, Mt. Lawley, W.A.

"   "   . "Gold Vol. 2: A Guide to Alluvial and Dryblowing Patches in the Murchison-Peak Hill Region". T.E. Johnston and Associates, Mt. Lawley, W.A.

"   ", 1983. "Gold Vol.3: A Guide to Alluvial and Dryblowing Patches in the Ashburton, Pilbara and Kimberley Regions". T.E. Johnston and Associates, Mt. Lawley, W.A.

Prospecting Methodology:

Garland, H.K., 1982. "All About Prospecting". Reed, N.S.W., 111p.

Picture Credits
Cover pictures: Gibson Consulting
All other photos and diagrams: By the author

www.ingramcontent.com/pod-product-compliance
Lightning Source LLC
Chambersburg PA
CBHW062029210326
41519CB00060B/7364